Walid Kaakeh

The Thinking Green: Unlocking the Secret Minds of Plants

Walid Kaakeh

The Thinking Green: Unlocking the Secret Minds of Plants

How They Perceive, make Decisions, and Respond Without a Brain

Noor Publishing

Imprint
Any brand names and product names mentioned in this book are subject to trademark, brand or patent protection and are trademarks or registered trademarks of their respective holders. The use of brand names, product names, common names, trade names, product descriptions etc. even without a particular marking in this work is in no way to be construed to mean that such names may be regarded as unrestricted in respect of trademark and brand protection legislation and could thus be used by anyone.

Cover image: www.ingimage.com

Publisher:
Noor Publishing
is a trademark of
Dodo Books Indian Ocean Ltd. and OmniScriptum S.R.L publishing group

120 High Road, East Finchley, London, N2 9ED, United Kingdom
Str. Armeneasca 28/1, office 1, Chisinau MD-2012, Republic of Moldova, Europe
Managing Directors: Ieva Konstantinova, Victoria Ursu
info@omniscriptum.com

Printed at: see last page
ISBN: 978-620-8-87065-2

The Thinking Green: Unlocking the Secret Minds of Plants

How They Perceive, Make Decisions, and Respond Without a Brain

Prof. Dr. Walid Kaakeh

Agricultural Consultant & Expert

بسم الله الرحمن الرحيم

IN THE NAME OF ALLAH

THE MOST BLESSED, THE MOST MERCIFUL

Dedication

To my mother and father,
Aisha and Abdul Ghani Kaakeh
May God bless them and make them dwell
in His spacious gardens

To my wife
Fatima Khatib
for her unwavering patience, compassion, and most
importantly her love

To my beloved children,
Tasneem, Yasmine, Yamen, and Enas
who continue to be a source of inspiration and pride

I dedicate this book

Walid Kaakeh

Table of Contents

Abstract

The Thinking Green: Unlocking the Secret Minds of Plants explores the provocative idea that plants—though lacking brains or neurons—demonstrate complex behaviors traditionally associated with intelligence. Drawing on cutting-edge research in plant neurobiology, ecology, and Indigenous knowledge, this book examines how plants perceive their environment, respond to challenges, remember past experiences, and communicate with one another.

Through vivid storytelling and scientific depth, readers will journey from the roots to the canopy, uncovering how plants "see" light, "hear" vibrations, and "speak" in chemical signals. Chapters delve into symbiotic relationships, plant memory, communal networks, and the subtle rhythms of growth and response that defy our human-centric notions of cognition.

More than a scientific exposé, this is a philosophical invitation to rethink life itself. What if intelligence is not confined to brains, but is distributed across living systems? What can the green world teach us about resilience, cooperation, and harmony?

The Thinking Green challenges our assumptions and expands our minds—through the living wisdom of plants.

Preface

We walk beneath trees and tread over mossy roots, often unaware that the world around us is pulsing with perception, response, and memory. Plants are the Earth's oldest inhabitants—silent, rooted, and often dismissed as passive fixtures of our environment. But what if that stillness conceals a sophisticated, intelligent life that communicates, learns, and adapts in ways we are only beginning to comprehend?

This book was born from wonder—a curiosity about whether the green life we take for granted might possess something akin to consciousness. Could plants, in their quiet way, *think*? Could they perceive the world not just as stimuli to react to, but as an intricate, interconnected network through which they respond with agency, purpose, and even memory?

Scientific exploration is increasingly revealing that plants do far more than photosynthesize and grow. They detect light, sound, touch, chemicals, and gravity with astounding sensitivity. They engage in symbiotic relationships, warn their kin of danger, form alliances with fungi, and even change their behavior in response to prior experiences. Far from passive beings, they are problem-solvers, strategists, and survivors.

In the pages that follow, I invite you to walk through forests not just as a hiker, but as a fellow participant in an intelligent conversation. This is not a work of fantasy, but one rooted in credible, emerging plant neurobiology, ecological science, and Indigenous wisdom traditions. Where science meets story, we begin to uncover a deeper truth—that intelligence is not limited to creatures with brains and neurons. It is distributed, embedded in life itself, and flourishing in the green silence around us.

This book is for anyone who has ever marveled at the spiral of a tendril, the patience of a tree, or the wisdom in a garden. It is an attempt to shift our thinking, to cultivate a reverence for plants not as background scenery, but as co-inhabitants of the Earth—vital, aware, and full of potential insight for how we might live better.

Let us begin, then, not by asking *do plants think like us*, but rather, *what kind of thinking is this, and what can it teach us?*

Introduction

Rethinking Intelligence Without a Brain

The prevailing narrative of human history has long placed thought and awareness within the confines of a skull. Intelligence, we believed, required a brain—a centralized processor, a command center of neurons firing in complex patterns. But what if this assumption has limited our understanding of life's full potential? What if, by looking beyond the brain, we discovered intelligence not as an organ, but as a process distributed throughout a being?

Welcome to the world of plant cognition.

In this book, we will explore the growing field of plant neurobiology—not the literal presence of neurons or synapses in flora, but the recognition that plants perform cognitive-like tasks. They sense, process, communicate, and even "remember" in ways that suggest a decentralized yet highly intelligent system.

Modern science is just beginning to catch up to what Indigenous cultures, naturalists, and poets have long known: plants are sentient in their own way. They do not think like us, but they think *nonetheless*—through chemical pathways, electrical signals, hormonal pulses, and even epigenetic memories passed through generations.

We'll begin by exploring how plants "see" without eyes, "hear" without ears, and "smell" without noses. We'll look at the mechanisms behind their ability to perceive the environment with profound acuity—from following the sun to sensing gravity and even responding to vibrations and touch. These senses, while alien to our own, allow plants to act with intention and adaptiveness.

Next, we'll explore the vast communication networks among plants—via mycorrhizal fungi, volatile chemical signals, and root-to-root contact. We'll consider whether these conversations are strategic, even social, in nature. We'll also examine memory and learning in plants, drawing from recent studies that show how plants change their behavior based on past experiences.

Beyond the laboratory, we'll journey into forests, deserts, and fields, examining the broader ecological intelligence of plant communities. What lessons can we learn about cooperation, resilience, and balance from organisms that have flourished for millions of years without domination or haste?

Finally, we will consider the ethical implications of plant awareness. If plants are intelligent, how might this reshape our agriculture, medicine, conservation, and even our spirituality?

This is not a book that anthropomorphizes plants. It does not imagine them to be miniature humans, thinking in words or planning in concepts. Instead, it invites us to expand our definitions of thought and consciousness to accommodate forms that are quiet, rooted, and ancient. In doing so, we may discover a more connected, humble, and reciprocal relationship with the living world.

Chapter 1
Light Watchers: How Plants See Without Eyes

"There are none so watchful as those who appear still."

Introduction

In the tranquil stillness of the early morning, as the first rays of sunlight filter through the landscape, an ancient and silent drama unfolds. Along a rolling Tuscan hillside, rows of sunflowers rise like sentinels, their golden faces stretching towards the warmth of the sun. They appear still, unmoving, and yet they are in the midst of a meticulous, synchronized ritual. No eyes, no muscles, no neurons—yet they "see." The plants do not require the complex sensory apparatus that humans or animals use to perceive the world. Instead, their existence is governed by the subtleties of light, which they can perceive with incredible precision.

For centuries, we believed that plants were fundamentally passive organisms, incapable of perception in the same way animals are. Fixed in place, they appeared to simply grow and respond to environmental conditions without any awareness of the world around them. But as science has revealed, plants possess a sophisticated system for perceiving and reacting to light—a system that opens a window into a world we once thought them blind to.

1.1. The Invisible Eyes of the Green World

Unlike animals, which have eyes to process visual stimuli, plants do not "see" the world in images. But their perception of light is no less extraordinary. At the core of plant light perception are photoreceptors—specialized proteins that respond to different wavelengths of light. These molecular sensors are attuned to specific regions of the light spectrum, including blue, red, and far-red light. Among the most important of these receptors are phytochromes, which respond to red and far-red light, and cryptochromes, which are sensitive to blue and ultraviolet light.

These photoreceptors enable plants to detect light quality, direction, intensity, and even the duration of exposure. For example, a seed lying just beneath the soil can sense the type of light filtering through the earth above. If it detects an absence of light—indicating a blockage, such as a rock or another plant—it will refrain from germinating. However, if the seed detects an open space, it will sprout and grow towards the surface.

Moreover, plants can sense the ratio of red to far-red light, a phenomenon that provides them with critical information about the presence of nearby plants. If the ratio is high, it suggests an open area with minimal competition for sunlight. If the ratio is low, it means that another plant is nearby, casting shadow and competing for resources. Armed with this knowledge, the plant adjusts its growth—bending towards light or stretching to escape the shade—engaging in what is known as shade-avoidance behavior.

In essence, plants are continuously "watching" their environment and responding accordingly. Their perception is finely tuned to the dynamics of light, guiding them in their search for survival, growth, and reproduction.

1.2. Memory Without a Brain?

Plants do not have brains, yet they exhibit behavior that suggests some form of memory. Consider the movement of chloroplasts—those specialized organelles responsible for photosynthesis. Chloroplasts are capable of moving within plant cells, repositioning themselves to capture the optimal amount of light. This movement is dynamic and responsive, allowing the plant to adapt to changing light conditions.

What is even more intriguing is the plant's ability to "remember" past light exposures and adjust its future behavior accordingly. Certain plants, such as ferns and mosses, retain a record of previous light conditions. This memory does not involve neurons or synapses but is encoded in cellular processes, possibly through changes in calcium ion channels or shifts in gene expression.

This ability to "remember" past light exposures represents a primitive form of memory, wherein the plant learns from past experiences and adapts its future responses. Although this memory does not resemble the neural memory of animals, it allows plants to optimize their behavior in response to light—a fundamental aspect of survival.

1.3. The Case of the Sunflower

Among the most well-known examples of plant light perception is the sunflower, which displays a fascinating phenomenon known as heliotropism. Young sunflowers track the movement of the sun across the sky, turning their heads from east to west as the day progresses. This movement is not a random or passive response; it is a carefully coordinated process driven by the sunflower's internal circadian rhythm.

The sunflower's ability to anticipate the sun's position is not a simple reaction to external stimuli. It involves the synchronization of its internal biological clock with the movement of the sun, adjusting its cellular and hormonal activity to facilitate optimal growth. During the night, the sunflower resets its position in preparation for the next day's cycle.

However, this movement ceases in mature sunflowers. Instead of following the sun, the mature flower remains fixed in a permanent eastward-facing position. Why east? Because morning sunlight attracts pollinators, increasing the chances of successful reproduction. By facing east, the sunflower optimizes its chances of ecological success—not only by maximizing its exposure to light but also by enhancing its ability to interact with its environment.

1.4. Seeing Without Seeing

The idea that plants can "see" without eyes may seem paradoxical, but the evidence is clear: plants respond to light with remarkable precision and sophistication. When plants are deprived of their photoreceptors, they behave as if they are blind—unable to track the sun, orient themselves to light, or effectively compete with other plants. Without their ability to sense light, plants cannot thrive in their environments.

However, in the wild, plants survive and even flourish by using light-based intelligence, relying on mechanisms far different from the visual perception systems of animals. The key to understanding plant "sight" lies not in the eyes but in the act of detecting and interpreting light to guide behavior. This form of perception allows plants to "see" their environment, navigate challenges, and make decisions based on light cues.

In this sense, the act of seeing extends beyond the traditional concept of vision. It is a broader, more inclusive idea that encompasses the ability to detect and respond to environmental stimuli—whether those stimuli come in the form of light, sound, or other cues. For plants, light is the primary source of information, and it shapes their behavior in profound and often unseen ways.

1.5. Green Thought

"We imagine intelligence through mirrors of ourselves—yet the leaves have always watched us in silence."

As humans, we often define intelligence through mirrors of ourselves, imagining that consciousness and awareness must operate through the same structures and processes we use. But perhaps this assumption has blinded us to the intelligence that exists in other forms of life. The leaves of plants, silent and still, have always "watched" us in their own way, attuned to the rhythms of light, growth, and survival. They may not see as we do, but in their subtle interactions with light, they demonstrate a form of awareness that is both ancient and profound.

In this light, we are challenged to reconsider what it means to "see" and to "know." Perhaps, as we begin to understand the intricate ways in which plants perceive their world, we will recognize that intelligence is not solely defined by eyes, brains, or synapses, but by the capacity to sense, respond, and adapt to the environment—no matter how still or silent one may seem.

This chapter invites us to look deeper into the world of plants, recognizing that their perception of light offers a fascinating and complex form of sight—a sight that transcends the traditional boundaries of vision and opens a new lens through which we can understand life itself. Through light, plants watch the world and shape their existence, teaching us to expand our definition of awareness and intelligence.

Chapter 2

The Roots of Hearing: How Plants Listen to the World

"Silence is not the absence of sound—it is the space where listening begins."

Introduction

In a quiet laboratory in Florence, Italy, a row of pea seedlings trembles ever so faintly in the dark. There is no breeze, no touch, and no obvious stimulus. And yet, beneath the surface, something ancient is stirring. A low-frequency hum vibrates through the air—recordings of water flowing through pipes—and the plants, drawn like diviners to a hidden spring, begin to angle their roots in its direction.

What does it mean for a root to listen?

This is the paradox that researchers like Dr. Monica Gagliano dared to ask when most botanists would have dismissed the question outright. Hearing, after all, was long considered a sense reserved for animals with ears and eardrums. But Gagliano's experiments cracked open the old assumptions. In carefully controlled studies, she demonstrated that plant roots could not only detect sound vibrations but also distinguish between them—growing preferentially toward the frequency associated with running water, even when no moisture was present. This is not magic. It is not mysticism. It is biology we have yet to fully understand.

Let's explore how plants use sound, vibration, and frequency as vital sensory inputs, expanding our understanding of plant life in ways we had never imagined.

2.1. The Symphony of Vibration

Sound, in its most basic form, is a vibration in a medium—usually air or water. These vibrations can be felt as much as heard, and at certain frequencies, even the smallest cells can perceive the tremors.

For plants, these vibrations aren't just passive disturbances—they are essential cues to the environment. Roots are exquisitely sensitive to mechanical signals. The soil they inhabit crackles with microscopic pressure changes, and tremors ripple through it from footfalls, wind, and the movement of insects. Their very survival depends on detecting the hardness of rocks, the softness of loam, and the presence of moisture hidden deep in the earth.

Tiny mechanoreceptors within root cells act like biological microphones, translating vibrations into chemical messages that guide their growth. These receptors are embedded in the cell walls, specialized to detect the subtlest vibrations, and they are attuned to different frequencies of sound, from the very low (in the range of 10 Hz to 220 Hz) to higher-frequency vibrations. When a plant root encounters vibrations—whether from water flowing underground or the soil's shift due to wind—it can "feel" the presence of these signals. This information helps guide its growth direction, prompting the root to navigate toward or away from particular stimuli, such as moisture or obstacles like rocks.

One landmark study showed that maize roots emit clicking sounds—brief pulses at around 220 Hz—and other nearby roots responded to those clicks by changing direction. It is as though roots are not only listening but speaking, creating a hidden sonic language below the soil. The idea of plant echolocation—using sound and vibration to sense the environment—has emerged as a possible explanation, though it remains an area of intense research.

What does this suggest? It hints that plant roots may have evolved an intricate form of communication through vibration, which allows them to function in harmony with their surroundings. The ability to sense water flows, detect the presence of other roots, or even sense the approaching footfalls of large animals could be crucial for survival, growth, and resource allocation. This emerging form of "sound intelligence" is an area ripe for exploration and discovery, one that challenges the conventional view of plants as silent, passive organisms.

2.2. Plants and Music: Beyond the Myth

The idea that plants "like music" has long been a part of popular folklore. Many a gardener has played Bach to their begonias or whispered sweet nothings to their spider plants. While these claims have often been dismissed as fanciful thinking, science is beginning to uncover evidence that sound, and even music, does indeed influence plant growth.

In several studies, plants exposed to harmonious classical music, particularly sounds in the frequency range associated with soothing harmonics (around 100–500 Hz), grew taller and healthier than those subjected to discordant sounds—or to silence. These studies suggest that certain frequencies stimulate gene expression related to growth and defense. On the other hand, discordant or chaotic vibrations could inhibit plant health, perhaps mimicking stress or distress signals.

But it's not just about "pleasant" music. Specific frequencies seem to prime plants for various environmental conditions. For example, plants can detect the vibrations caused by insects chewing on their leaves. When a caterpillar begins to feed, the plant can "hear" the crunching sound of its jaws. This auditory stimulus triggers the plant to produce bitter chemicals in its tissues, effectively "warning" it of a potential predator.

Imagine a leaf that listens not for the approach of an insect via sight or smell, but by tuning into the vibration caused by its movements. In this case, the plant's response to sound seems to enhance its defense mechanism before the actual physical threat occurs. This adds an entirely new layer to the concept of plant intelligence, suggesting that plants are not only capable of responding to sound but are also preemptively preparing for the future.

2.3. Rethinking Ears and Empathy

This chapter isn't about anthropomorphizing plants, attributing human-like qualities to them, or imagining that they "feel" in the way humans do. Instead, it's about decentering ourselves from the concept of hearing. Hearing, like seeing, is a form of perception. It's not necessarily about the physical presence of ears or auditory organs, but about the ability to detect patterns in energy and vibration. Plants do this with incredible elegance and efficiency. Without brains, they decode the language of sound that pulses through the soil, the air, and even the rain. Plants are tuned into the sounds of their environment in a way that allows them to thrive.

In their world, the sounds are not always obvious to our ears, but they are present. The whistling wind, the rustling of leaves, the drone of insects—all of these sounds, though silent to us, are vital signals in the plant kingdom. Plants listen, not to hear in the traditional sense, but to interpret the vibrations that signify the presence of water, danger, or the changing environment. In essence, they are attuned to the hidden symphony of the natural world—a world alive with whispers, creaks, hums, and songs.

2.4 Green Thought:

"Perhaps the world is full of listeners—but we've mistaken stillness for silence."

In this moment of reflection, we are asked to reconsider our perceptions of stillness. What seems like silence is, in fact, a cacophony of vibrations and sounds, each one significant in the plant world. We have too long assumed that silence is the absence of noise—an empty space in which nothing occurs. But for plants, silence is a moment of profound listening, a time when they absorb all the subtle vibrations in the air and soil.

This chapter challenges us to rethink the very idea of hearing and perception. It's not merely about having ears; it's about detecting the world's invisible rhythms and responding to them. Whether through the vibrations of sound, the patterns of light, or the subtle shifts in their environment, plants are constantly aware, constantly learning, and constantly interacting with the world in ways that we are just beginning to understand.

As we delve deeper into the remarkable abilities of plants, we see that their silent wisdom offers us a profound lesson: Perception is not confined to the senses we know. It's a dynamic, ever-present force that plants have mastered long before we ever recognized it. In their stillness, they are always listening, always perceiving, and always responding.

Chapter 3

The Silent Language of Chemicals: How Plants Smell and Taste

"If you have ever walked through a forest, you know that there are more conversations happening than you can ever understand."

Introduction

Imagine standing beneath an ancient oak, feeling the breeze carry the scents of wildflowers across the land. To humans, the air is filled with a fragrant, beautiful aroma, perhaps even a fleeting memory. But for the oak, this same breeze carries much more—an intricate and deliberate message encoded in chemicals, signaling to the tree something profound. Perhaps it's a warning of danger, an invitation to form a symbiotic relationship, or a cue that the seasons are shifting. This invisible message isn't meant for the human nose; it's designed for the oak and other plants within its reach, signaling chemical information that guides responses crucial for survival.

Plants, as it turns out, are experts at using chemicals to communicate with each other and the world around them. What we perceive as "smell" is not a passive response to the environment; it is a dynamic, active form of interaction, woven into the biology of plants. They communicate with molecules known as volatile organic compounds (VOCs), using scent as a language. VOCs are not random emissions or by-products of metabolism—they are strategic signals that can alter behavior, protect against threats, and even influence the growth of neighboring plants.

3.1. The Language of VOCs: A Symphony of Scents

Imagine being a plant. Stuck in one place for your entire life, you might seem like a passive observer, waiting for something to happen. But that couldn't be further from the truth. Plants are constantly engaged with their environment through a sophisticated chemical dialogue. This invisible network of communication allows them to convey their needs, fears, and desires to other plants, animals, and even microorganisms in their surroundings.

When a plant is under attack by herbivores, it does not fight back by screaming or physically defending itself. Instead, it releases VOCs into the air—chemical messages that can travel long distances, informing nearby plants of danger. These VOCs serve as a warning to other plants to "brace themselves" and bolster their defenses. Sometimes, these signals also attract natural predators to the herbivores, such as parasitic wasps that hunt caterpillars or ladybugs that feast on aphids. The VOCs work like an alarm bell, but with a subtle, often undetected force.

For instance, if a caterpillar begins munching on a plant, the plant responds to the vibrations and the damage by releasing a cocktail of chemicals that acts as both a deterrent and an invitation to predators. The plant's "smell" doesn't just warn other plants—it sends out a message that can shift the dynamics of the entire ecosystem.

3.2. The Secret Life of the Wood Wide Web

Beneath our feet, within the soil, a hidden network of communication exists, often referred to as the "Wood Wide Web." This subterranean network is a complex system of mycorrhizal fungi that links the roots of various plants and trees. It is through this underground system that plants exchange vital information about their environment, from nutrient availability to threats of disease or insect invasions.

The mycorrhizal fungi act as messengers, carrying chemical signals between plants. For example, when one tree is attacked by a pest or is struggling to find nutrients, it can send chemical distress signals through this network to its neighbors. These neighboring plants then prepare their defenses or adjust their growth patterns in response. In some cases, trees may even "share" resources, such as water or nutrients, when one is suffering from drought or disease, helping ensure mutual survival.

This cooperation isn't an act of selflessness; it's a strategic move for mutual benefit. The plants are essentially sharing a network of resources and information to ensure their collective survival. This "chemical democracy" is a form of symbiosis, where the success of one plant contributes to the success of the larger ecosystem.

3.3. Carnivorous Plants: When Smell Becomes a Trap

Some plants take chemical signaling to an entirely different level, using scent as a means to trap and consume animals. Carnivorous plants like the Venus flytrap and the pitcher plant use scent not as a mere signal, but as a bait for unsuspecting prey.

The Venus flytrap, for example, emits a sweet, enticing fragrance that draws insects toward it. This scent, rich in nectar-like chemicals, is a signal to the insect that it has found a safe and nourishing place. But the moment the insect lands on the trap, the plant snaps shut, trapping the prey inside. The plant then digests its catch, drawing nutrients from the once-living insect to fuel its growth.

Similarly, the pitcher plant releases a scent that mimics the smell of decaying flesh. This carrion-like odor is irresistible to certain species of flies and other insects that are naturally attracted to rotting meat. The insects fly toward the pitcher plant's bulbous trap, slip into its slippery interior, and fall into a pool of digestive enzymes. The plant uses scent as both an invitation and a trap, expertly turning its chemical signals into a form of predation.

3.4. The Symbiosis of Scent and Survival

The role of scent in plant survival extends far beyond the realms of attraction or defense. It's not just about flowers calling to pollinators or trees warning their neighbors of danger. Scent, in the plant world, is a crucial tool for maintaining and enhancing survival within complex ecosystems. For example, when a tree releases VOCs signaling distress, it doesn't only alert its immediate neighbors—it can also signal a larger network of plants in the area, which may then adjust their growth or defensive tactics.

The release of these compounds can change the behavior of animals and insects. Pollinators, like bees, may be drawn to a particular flower's fragrance, but in the same breath, that same scent may signal to herbivores to stay away. Plants constantly measure and adjust the chemical signals they send out to maintain a delicate balance of attraction and repulsion.

Moreover, this network of scent-based communication links different species in an ongoing dance of survival. When a plant releases a certain compound to attract a pollinator, it is simultaneously affecting the surrounding flora and fauna. The entire ecosystem operates through this web of chemical signals that guide not only plant growth but also animal behavior, creating a highly synchronized, albeit unseen, symbiosis.

3.5. Green Thought

"We are not alone in this world of words; we simply need to learn to listen to the language of the leaves and the roots."

In the silent world of plants, language is alive and constantly shifting. It's a language that we, as humans, are only beginning to uncover. Plants communicate not only through what we perceive as smell and taste but through a series of sophisticated chemical signals that navigate the air, soil, and even the spaces in between. By learning to listen and understand the silent messages of plants, we open a window into a world that is teeming with stories, strategies, and relationships we've only begun to comprehend.

Through chemical communication, plants are not mere passive organisms reacting to their environment—they are active participants in a larger, intricate conversation that spans ecosystems, connecting them in ways that are both subtle and profound. The next time you encounter a flower or pass through a forest, know that the air is thick with stories, with signals and responses that are more complex and beautiful than you can ever hear. The scent of a plant is never just fragrance—it's a conversation, a call to action, and a map of survival.

Chapter 4

Touch and Memory: How Plants Feel and Remember

"In the language of the trees, silence is not absence; it is the echo of touch, the memory of light."

Introduction

Touch—the word itself invokes images of physical closeness: a hand brushing against a cheek, a hug that envelops. Yet, in the plant kingdom, the concept of touch transcends physical contact as humans understand it. Plants, despite being immobile, engage in a delicate and multifaceted dialogue with the world around them. Their sensory capabilities, particularly touch, are not just vital for survival, but a key to their evolution, adaptation, and communication. Plants, though rooted, are far from passive entities. They respond to tactile stimuli from the environment, sensing the pressure of the wind, the caress of an insect, or the disturbance caused by a passing herbivore. These sensations trigger chemical responses, adaptive behaviors, and even a kind of memory that helps plants navigate the complex and often hostile landscapes they inhabit.

In this chapter, we explore the intricate ways plants perceive touch, how they "remember" experiences, and how these memories influence their survival and interaction with their environment. The concept of "memory" in plants, though different from that of animals or humans, plays a profound role in shaping their responses to the world.

4.1. The Whisper of the Wind: How Plants Sense Touch

While humans rely heavily on sight and sound, plants have evolved a different set of sensory mechanisms to perceive and respond to their environment. Mechanoreceptors, specialized cells designed to detect physical forces, are at the core of this unique sensory system. These receptors allow plants to sense even the subtlest of touches—from the gentle breeze to the heavy force of an herbivore grazing.

One of the most fascinating examples of plant touch sensitivity is seen in the *Mimosa pudica*, often called the "touch-me-not" plant. When its leaves are brushed, they rapidly fold inwards, seemingly recoiling in response. While this may appear to be a dramatic display, it is, in fact, a defensive tactic. The plant reacts in this manner to discourage herbivores from further grazing. This instantaneous movement demonstrates not only the plant's ability to "feel" touch but also its ability to respond swiftly and appropriately to perceived threats.

Likewise, climbing plants, such as the pea plant (*Pisum sativum*), rely on their touch-based memory for survival. Their tendrils, upon making contact with a nearby structure, immediately curl around it. This allows the plant to secure itself to a support, ensuring its growth is not impeded. However, this "memory" goes beyond simple mechanical movement. These tendrils have a form of learned behavior—they "remember" the touch of a surface and use this information to direct their growth in search of a suitable structure to grasp. The plant does not simply react; it actively learns and adapts based on tactile feedback.

Through these examples, it becomes clear that plants' sense of touch is not passive. It is an active and adaptive process that influences their growth, behavior, and survival.

4.2. Memory: How Plants Remember the Past

Memory, often understood as the ability to store and recall past experiences, seems like a distinctly human trait. Yet, plants exhibit a form of memory, though it operates in a fundamentally different way. Without brains or central nervous systems, plants do not store memories in the way animals do. Instead, their memories are encoded in chemical responses and physical changes that are shaped by past experiences.

Plants' memory is most apparent in their ability to respond to environmental stresses. For example, consider a plant that experiences a period of drought. The plant will undergo biochemical changes to conserve water, such as closing its stomata (pores on the leaves) to reduce water loss. However, after this drought has passed, the plant "remembers" the experience through biochemical and physiological adjustments. For the next drought, the plant might be more efficient in conserving water or even develop deeper roots to access moisture buried deeper in the soil.

This ability to remember past stressors and adapt is referred to as "priming." It is a type of biological memory that enhances a plant's capacity to respond to recurring environmental challenges. Furthermore, when a plant faces a pathogen, it mounts an immune response. Interestingly, if the same plant is exposed to the same pathogen again, it will often mount a stronger and faster defense. In a communal sense, plants can also share these memories with neighboring plants. Through chemical signaling, they can warn and "inform" other plants of potential threats, ensuring that the whole ecosystem is better prepared for future challenges. This sharing of memory and knowledge is a form of collective intelligence that is crucial for the survival of plant communities.

4.3. The Movement of Memory: A Plant's Experience of Time

The way plants experience time is distinct from the human perception of past, present, and future. While humans view time in a linear fashion, plants experience it more cyclically, as they grow and adapt to the changing seasons and environmental shifts. Time, for plants, is recorded not in minutes or hours, but in the growth cycles and seasonal rhythms that shape their lives.

Consider the rings of a tree. Each year, a tree produces a new layer of wood, forming a growth ring. These rings hold within them the memory of the conditions the tree endured during that year. A wide ring may indicate a year of abundant rainfall, while a narrow ring might suggest a year of drought. Through these rings, trees leave behind a record of their experiences, an autobiography written in wood. Each growth ring is a chapter in the tree's life, chronicling its survival through the seasons.

Similarly, plants also possess an internal clock, which is regulated by light cycles. This is known as photoperiodism, and it is a mechanism by which plants "remember" the length of days and nights. This memory of light cycles dictates critical processes such as flowering, fruiting, and dormancy. In this sense, plants are attuned to the passage of time in a way that allows them to synchronize their biological functions with the changing seasons. Their memory of time is encoded not in the form of conscious recollection, but in the very rhythms of life.

4.4. The Touch of Memory and Survival

Memory, in the plant kingdom, is not confined to the individual. It extends across species, communities, and ecosystems. The memories of individual plants contribute to the collective survival of the entire species, shaping the dynamics of plant communities and the environment as a whole.

Each touch—a gust of wind, the brushing of an insect's antenna, or the nibbling of a herbivore—leaves a lasting imprint. These tactile experiences shape the plant's responses, adjusting its growth patterns, chemical defenses, and overall behavior. Touch, for plants, is not a passive experience; it is a profound interaction that influences their survival and adaptation over time.

Plants also "remember" the touch of other plants. In the face of damage or stress, plants may release volatile organic compounds (VOCs) that signal neighboring plants to prepare for similar challenges. These VOCs serve as chemical warnings, alerts that the ecosystem needs to brace for potential threats. This type of communication ensures that plants are not isolated in their struggles but are part of a larger, interconnected network.

As with human memory, the touch of the world shapes the plant's identity, its survival strategy, and its relationship with its environment. The world around them—full of tactile sensations—forms the backdrop against which plants thrive, adapt, and evolve.

In the end, memory and touch in plants are not abstract or passive concepts. They are essential components of the plant's interaction with the world, allowing them to survive, adapt, and communicate in ways that we are only beginning to understand. Each touch, each encounter, shapes the plant's future, just as it shapes the world in which it exists.

Chapter 5
The Heartbeat of the Earth: How Plants Move and Respond

"The roots do not just anchor the tree; they listen to the rhythm of the earth's pulse, responding to its every call, its every whisper."

Introduction

Movement, when we think of it, is usually a characteristic of animals and humans. It is dynamic, quick, and observable. We imagine plants to be passive, their only apparent movement being their steady growth. But this perception is both limited and misleading. The plant kingdom, as it turns out, is teeming with hidden, subtle, and profound movements—movements that respond to the world's rhythms, even though they may not be as immediate or noticeable as those of more mobile creatures.

The movement of plants is not defined by speed, but by the timing, adaptability, and precision with which they adjust to their surroundings. Unlike animals, plants do not move in frantic bursts, but in quiet, steady rhythms. These rhythms are driven by an intricate sensitivity to the Earth's pulse—its seasons, its subtle shifts, and its energy patterns. Through this responsiveness, plants connect deeply with the world, engaging in a complex dance that is barely perceptible but immensely significant.

The true essence of plant movement lies not in its speed, but in its methodical, patient, and profoundly adaptive nature. And as we begin to unravel the subtleties of this movement, we begin to understand that plants are far from stationary. They are in constant dialogue with the world, responding to every change with an attentiveness and wisdom that we are only beginning to appreciate.

5.1. The Silent Dance – Roots and Their Invisible Movements

Below the surface of the soil, the roots are alive with movement. We may not be able to observe it, but their activity is ceaseless and dynamic. The roots grow, spread, and respond to environmental cues in ways that challenge our conventional understanding of plant life.

Roots are not simply passive structures that anchor plants into the soil. They are astutely aware of their environment, moving with purpose as they respond to gravity, moisture, and the composition of the soil. The notion of roots "searching" for water, nutrients, and space is not merely a metaphor—it is a reality. Roots actively seek these resources, weaving their way through soil, stones, and other roots, adapting their path based on what they encounter.

At first, a newly emerging root seems to grow in random directions. It encounters obstacles, and rather than simply pushing through, it adapts, bends, and reorients. This behavior is not a random process but a learned response to the tactile feedback it receives. Roots "remember" the landscape they navigate, avoiding previously encountered barriers and optimizing their growth towards favorable conditions. They are not simply growing, but responding to the unique, ever-changing dynamics of the soil.

The complexity of these movements becomes more apparent when we consider how roots form intricate, interconnected networks. Through these networks, roots communicate with one another, sharing resources and information about the surrounding environment. This interaction is a kind of memory—a deep, ancestral knowledge passed from one plant to another, ensuring that each root knows where to go and how to thrive.

5.2. The Pulse of the Plant – Circadian Rhythms and Growth

Just as the roots communicate with the soil, the plant's above-ground structures engage in a silent dialogue with the cosmos. The most visible example of this is the movement of leaves, which follows a rhythm in sync with the cycles of day and night.

Plants are in tune with time—specifically, their movements are governed by circadian rhythms, the internal biological clocks that regulate many of their processes. This natural timing ensures that plants grow and respond to their environment in harmony with the broader cycles of the Earth. During the day, plants open their leaves to absorb sunlight for photosynthesis. As night falls, they close, conserving energy and resting until the next cycle. This movement, although slow and subtle, is a direct response to the ebb and flow of the environment around them.

Like the tides that rise and fall under the influence of the moon, the movement of leaves is guided by the rhythms of light and dark. Plants, in their quiet wisdom, know when to absorb energy and when to rest. This is not simply an adaptation to light—it is an expression of their sensitivity to the broader rhythms of existence. Their movements reflect an understanding of time, an awareness of their place within the intricate dance of life on Earth.

5.3. The Quiet Strength of Thigmotropism – How Plants Respond to Touch

Thigmotropism is another form of plant movement, one that occurs in response to touch. While most movements in the plant kingdom seem rooted in growth patterns or reactions to environmental factors, thigmotropism emphasizes the plant's sensitivity to physical contact. This response plays a pivotal role in a plant's ability to interact with its environment.

A perfect example of thigmotropism can be found in climbing plants like vines or peas. These plants have specialized tendrils that, upon contact with a surface, wrap around it in a spiraling motion. This isn't simply an instinct to seek support—it is an active, deliberate movement that demonstrates the plant's ability to "feel" its surroundings and adjust accordingly. The tendrils do not just grow in any direction; they sense the object they encounter and respond by curling around it, using it for support as they continue their upward journey.

Thigmotropism also serves a defensive role. The sensitive plant, *Mimosa pudica*, exhibits a dramatic reaction to touch by folding its leaves in response to disturbances. This movement is more than just a defensive mechanism against herbivores; it is also a form of communication. By reacting in such a pronounced way, the plant signals to the world that it is not a passive target but an active, responsive organism.

5.4. The Movement of Life – How Plants Communicate Through Movement

Plants are not just passive entities—they are deeply communicative. Through their movements, they engage with their surroundings, sending signals and responding to external cues. These movements are often subtle, but they speak volumes about the plant's awareness of the world around it.

Take, for example, the slow, deliberate opening of a flower's petals in the morning. This movement is not just a response to light; it is an invitation. The flower is reaching out, sending a message to pollinators, signaling that it is ready to share its nectar and continue the cycle of life. This movement is not passive but a strategic, dynamic response to the environment's call. Similarly, the way a vine wraps around a support or the manner in which a sunflower follows the sun throughout the day—all these movements are ways in which plants communicate their needs and intentions.

In forests, trees communicate in quieter ways. Beneath the soil, through vast networks of roots and mycorrhizal fungi, trees exchange nutrients, warn each other of threats, and share resources in times of scarcity. This exchange represents a form of plant communication that is not just biological but a deep, interconnected web of mutual support. It is through these subtle exchanges that the movement of life flows, quietly and steadily, beneath the forest floor.

5.5. The Movements of the Earth and the Plants That Listen

The Earth itself is in constant motion. Its tectonic plates shift, its magnetic field pulses, and the very air we breathe circulates in invisible currents. Plants, in their deep connection to the Earth, respond to these movements in ways we are only beginning to understand.

Their roots stretch deeper during droughts, seeking water; their leaves turn toward the sun as it crosses the sky; their flowers bloom in synchrony with the seasons. Every movement plants make is a response to the forces of the Earth. And through these movements, plants continue to adapt, evolve, and thrive, proving that even in stillness, there is a profound response to the world's rhythms.

This quiet yet powerful movement reminds us that change, growth, and adaptation do not always require loud or rapid actions. Sometimes, it is the slow, patient movement of life that holds the most significance. Just as the Earth's pulse guides the plant, the plant's subtle movements reflect an understanding of the deep, rhythmic forces that shape all life.

In the end, the plant kingdom teaches us a valuable lesson: that movement is not always about speed or noise. It is about being present, adapting with patience, and responding to the silent, steady beat of the Earth's pulse. The plant's movement is not a race; it is a rhythm, a dance that connects it to the vast, interconnected web of life.

Chapter 6
The Eyes of the Forest: How Plants See and Respond to Light

"In the quiet of the forest, where shadows stretch and dance, the leaves listen not with ears, but with the very cells of their being. They see the light that caresses them, bending and shifting, following its every whisper."

Introduction

Light, that most elusive of forces, holds the world together. It is the unseen hand that tugs at the heart of all living things, compelling them to reach toward its embrace. The sun's rays, though they appear as mere beams of energy, are, in fact, the architects of life. Without light, there would be no warmth, no photosynthesis, no life as we know it. And in this light, the plants, though they cannot see in the way we do, have learned to listen, to feel, and to respond with a remarkable sensitivity that transcends our understanding of vision.

In the language of the forest, light speaks through the movement of leaves, the bending of stems, the opening of flowers, and the very growth of the plant. But this is not a passive response. Plants have developed an intricate, almost uncanny ability to "see" light in ways that are far beyond the human capacity for sight. They sense it, they orient to it, and they use it not only for growth but for survival. It is, in essence, their way of navigating the world, their compass in a universe of shadows and brightness.

6.1. The Whisper of the Sun – Photoreception and the Plant's Ability to Sense Light

To understand how plants "see" light, we must first dive into the world of photoreception—the process by which plants detect and respond to light. At the core of this ability are specialized proteins called photoreceptors, which are embedded in the cells of the plant. These proteins respond to different wavelengths of light, allowing the plant to perceive the sun's rays in a spectrum far beyond what our human eyes can detect. These photoreceptors, like silent sentinels, absorb the light's energy and transform it into a signal that the plant can understand, turning light into the very essence of growth.

In this intricate dance, plants are most attuned to the light that falls within the blue and red wavelengths of the spectrum. Blue light, with its high energy, tells the plant that it is time to begin photosynthesis, sparking the creation of sugars from sunlight. Red light, on the other hand, governs processes related to flowering and flowering cycles, directing the plant's energy to reproductive growth.

But there is more to light than mere illumination. It is a signal, a communication between the plant and the world around it. Just as the moon's soft glow moves the tides, the sunlight compels the plant to grow and stretch, shaping its form in response to every nuance in its environment.

6.2. The Movement Toward the Light – Phototropism and the Dance of Growth

Plants, unlike animals, cannot move in the traditional sense. They are rooted in place, but this rootedness does not limit their ability to respond to the world. In fact, plants have learned to move in the most subtle of ways, following the sun's path as it travels across the sky. This movement—this quiet dance toward the light—is called *phototropism*.

Phototropism occurs when plants grow toward light. The process begins when a plant detects the difference in light intensity between its shaded and sunlit sides. The plant's cells respond differently to light, elongating on the side that is less exposed to the sun, while compressing on the side that receives more light. This differential growth causes the plant to bend toward the light source, reaching for the sun with an urgency that belies its slow pace.

This ability is vital to the plant's survival, for it ensures that the plant maximizes its exposure to light, the source of energy that sustains it. A seedling, emerging from the soil, reaches upward instinctively, seeking the warmth and nourishment of sunlight. The plant's body—its leaves, its stems—are oriented toward the sun, growing ever higher, ever closer to the light that fuels its life.

6.3. The Art of Direction – The Role of Auxins in Light Sensing

At the heart of phototropism lies a complex molecular response involving a group of plant hormones known as *auxins*. Auxins are essential for regulating the plant's growth patterns, and they play a pivotal role in how a plant reacts to light. When a plant is exposed to light, the auxins accumulate on the shaded side of the plant, causing the cells to elongate and the plant to bend toward the light.

The brilliance of this mechanism lies in its simplicity and effectiveness. The plant does not need eyes or conscious thought to know where to grow. It is guided by a chemical signal, a messenger that speaks to it in the language of growth, telling the plant, "Move this way. Follow the light." This process is not only a remarkable feat of biological engineering but also a testament to the adaptability of life.

6.4. The Shade and the Sun – How Plants Compete for Light

While plants may reach toward light in a harmonious, graceful dance, not all interactions are peaceful. In the wild, plants must compete for access to light, and this struggle shapes their forms and their strategies for survival. In forests, jungles, and grasslands, where the canopy of trees casts dense shadows upon the forest floor, plants have adapted in various ways to ensure they capture the light they need.

Some plants grow taller, stretching upward in search of the sun, while others develop larger leaves to capture more light in their lower positions. The sunflower, for example, turns its head toward the sun, following its path across the sky with what appears to be a conscious intent. This behavior, called heliotropism, is a specific form of phototropism in which the plant orients itself to maximize sunlight absorption throughout the day.

But it is not just tall trees and reaching vines that compete for light. Beneath the surface, in the shadows, plants like ferns and mosses have learned to thrive with less. These "shade-loving" plants have evolved to make use of the limited light that filters through the forest canopy. Their leaves are adapted to absorb more of the sunlight that reaches them, making them highly efficient in their environment.

6.5. The Dance of Flowers – How Light Governs Reproduction

Beyond its role in growth, light plays a crucial part in the reproductive cycle of plants. The timing of flowering, the opening of buds, and the ripening of fruit are all governed by the rhythms of light. Plants are attuned not just to the presence of light, but to its duration—the length of day and night, the changing seasons, the shifting angles of the sun across the sky.

In many plants, the transition from vegetative growth to flowering is triggered by changes in the amount of light that the plant receives. Some plants are "short-day" plants, meaning they flower when the nights are longer than the days, while others are "long-day" plants, flowering when the days are longer. The flowering of many plants, including the iconic cherry blossoms of Japan, is timed with such precision that the entire community seems to bloom in unison, an orchestra of color responding to the same cosmic cue.

And in the night, flowers too respond to light—albeit in a different manner. Some flowers, like those of the evening primrose, open only when the sun has set, releasing their fragrance to attract nocturnal pollinators such as moths. It is a dance of opposites—the daytime flowers reach for the sun, while the nighttime flowers open to the moon's silvery glow. Each follows its own rhythm, dictated by light's silent force.

6.6. The Eyes That Never Close – The Persistence of Light Sensitivity

Though plants lack the complex eyes that animals possess, they are far from blind. Their sensitivity to light is an ever-present force, guiding them from the moment they emerge from the soil to the day they complete their life cycle. Even in darkness, the plant is listening, waiting for the light to return. It is as though the plant has its own internal clock, its own sense of time that keeps it aligned with the natural world.

This is why plants, though they do not have conscious vision, still "see" in ways that resonate with a deeper, almost mystical understanding of light's power. They bend, grow, flower, and fruit in response to a force they cannot resist—an eternal call to follow the light, to grow toward it, to become more than they were before.

In the light, plants find their true purpose, unfolding in perfect harmony with the world. They are not mere bystanders in the great cosmic dance, but active participants—forever reaching, forever responding, forever moving with the pulse of the sun.

Chapter 7

The Sound of the Forest – How Plants Respond to Sound and Vibrations

"In the silence of the forest, there are whispers that the leaves catch—vibrations in the air that speak a language older than words. Plants do not hear as we do, but they feel, they listen with the very pulse of their being."

Introduction

The forest is alive not just with the hum of life but with a symphony of sounds—subtle, invisible, yet profound. To the casual observer, these sounds are little more than background noise—perhaps the rustle of leaves in the wind, the call of a bird, or the hum of insects. Yet, beneath this veil of sound, the world of plants is more intricate than we ever imagined. While we, as humans, experience sound with our ears and complex auditory systems, plants perceive vibrations, the tremors that ripple through the air, soil, and water. Though they may not have the sense of hearing that we possess, plants have developed a way of "listening" to the vibrations that shape their environment.

This ability to sense and respond to sound and vibrations is one of the most mysterious aspects of plant life. Plants can detect the faintest tremors—whether they are the vibrations of wind, the footsteps of a passing animal, or the reverberations of sound waves traveling through the air. And they respond not with movement or noise, but with subtle changes in their growth, their behavior, and their very physiology. In essence, they listen to the world in ways that are beyond our comprehension, attuning themselves to the symphony of vibrations that surround them.

7.1. The Sound of the Wind – The Forest's Invisible Conversation

Consider, for a moment, the simple act of wind moving through a field or forest. We hear it as a gentle rustling or a powerful roar, depending on the force with which it blows. But for plants, this wind is not just a sound—it is a signal. The vibrations caused by the wind, as it flows through the leaves, branches, and stems of trees, send ripples of information throughout the plant. These vibrations influence how the plant grows, how it responds to external stimuli, and how it interacts with the world around it.

A study from the University of Western Australia discovered that when plants are exposed to the sound of wind, they exhibit changes in their physiology. Their stomata—the tiny pores on the surface of their leaves—open and close in response to the vibrations, allowing the plant to regulate its water loss more efficiently. This response is thought to be a survival mechanism, helping the plant to conserve water during times of stress. The wind, then, is not just a passive force; it is a constant companion, shaping the plant's life in ways we are only beginning to understand.

Yet, this response to wind is not just a passive adaptation. Plants, in their own quiet way, communicate through sound and vibration. When the wind blows, it sets in motion a complex series of signals that travel from leaf to root, from branch to flower, altering the plant's behavior in real-time. The plant listens to the sound of the wind and responds by adjusting its structure, its metabolism, and its very being, ensuring that it remains attuned to the ever-changing environment.

7.2. Vibrations in the Air – The Power of Sound Waves

Beyond the wind, plants can detect other sounds in their environment—sounds produced by animals, insects, or even the footsteps of a human. But what does it mean for a plant to "hear" these vibrations, and how does it affect their growth?

In the natural world, sound waves travel through the air as vibrations. These waves, though inaudible to human ears, carry information that is vital to plants. A study led by Monica Gagliano, a researcher at the University of Western Australia, revealed that plants are capable of responding to sounds in ways that suggest a deep sensitivity to vibrations. For instance, plants exposed to the sound of herbivores feeding on nearby plants showed signs of distress—leaves curled, growth slowed—indicating that they were "hearing" the sounds of potential danger. In response to these vibrations, the plants activated defense mechanisms, releasing chemicals to ward off herbivores or attracting predators that would deal with the threat.

This phenomenon suggests that plants are not merely passive recipients of environmental stimuli. They actively "listen" to the sounds of the world around them and adapt their behavior accordingly. In a sense, plants have evolved to be hyper-aware of the vibrations in their environment, using them as a means of communication, defense, and survival. They can "hear" the vibrations of predators, the calls of pollinators, and even the rumblings of nearby storms, preparing them for whatever challenges may arise.

7.3. The Role of Root Vibrations – How Plants Communicate Underground

While most of us focus on the interactions between plants and the air above ground, an equally important world exists beneath the soil. In this hidden realm, plants communicate and respond to vibrations that travel through the ground. Roots, unlike leaves, are especially attuned to the vibrations in the soil, where they form a network of intricate, unseen connections with other plants, fungi, and microbes.

When a plant senses vibrations in the soil, whether from the movement of animals, insects, or even the subtle shifts caused by the movements of other plants, it responds by altering its growth patterns. Some plants may grow their roots toward the source of the vibration, seeking nutrients or water, while others may retreat or shift their growth direction to avoid competition or danger.

Moreover, research has shown that plants can communicate through the vibrations in the soil. When one plant is attacked by herbivores, for instance, it may send out vibrational signals that other plants nearby can detect. These plants, in turn, may activate defense mechanisms, such as producing toxins or emitting volatile compounds to repel herbivores. This form of "underground communication" allows plants to work together, creating a more resilient ecosystem that responds to threats as a collective.

7.4. The Effects of Music – Do Plants Really Respond to Melody?

For centuries, there has been speculation about whether plants respond to music. The idea that music can influence plant growth dates back to ancient times, and in modern times, it has become a subject of debate. While scientific research has produced mixed results, there are fascinating studies that suggest plants are indeed sensitive to certain types of sound.

In experiments where plants were exposed to different types of music, some studies have reported increased growth in plants that were subjected to classical music, while plants exposed to heavy metal or rock music showed slower growth. The theory behind this phenomenon is that sound vibrations—specifically those at certain frequencies—can influence the movement of molecules and ions in plant cells, thereby affecting their metabolism and growth. It is as though the sound waves have a direct effect on the plant's internal rhythm, coaxing it into a state of heightened growth or stress depending on the nature of the vibrations.

However, it is important to approach this idea with caution, as the scientific community remains divided on the subject. While some studies suggest that plants may be able to respond to the vibrations of sound, others argue that the effects are more psychological than biological. Nevertheless, the concept of music and sound as a force that can influence plant life continues to intrigue and inspire curiosity.

7.5. The Quiet Intelligence of Plants – An Unseen World

The idea that plants can "hear" their environment challenges our fundamental assumptions about intelligence, perception, and communication. While we may never truly know the depth of their sensitivity to sound, vibrations, and the subtle energies of the world around them, what is clear is that plants are far more aware of their environment than we ever imagined.

In the silence of the forest, the plants listen—not with ears, but with roots, leaves, and cells that feel, vibrate, and respond to the invisible symphony of life. They "hear" the world in ways that transcend our understanding, using vibrations to navigate, communicate, and survive. In this quiet intelligence, plants reveal a hidden layer of consciousness, one that connects them to the pulse of the earth and the vibrations of the universe.

They may not speak, but in their stillness, they are listening. And in that listening, they are far more attuned to the world than we might ever comprehend.

Chapter 8

The Roots of Connection – How Plants Communicate and Form Social Bonds

"In the silence of the soil, beneath the canopy of leaves and branches, plants whisper to each other—through roots, through air, through shared chemical signals. They speak not in words, but in the language of life itself, a language older than speech, binding them in networks of unseen kinship."

Introduction

If we could only peer beneath the surface, into the realm where roots reach and intertwine, we would uncover a world of connection and communication far beyond our comprehension. The idea of plants as solitary beings—standing alone, silently growing towards the sun—is a simplistic view of nature's true dynamics. Beneath the soil, in the intricate web of roots and fungal threads, plants form vast, complex networks of interaction and communication, often working together in ways that mimic human society. They share resources, warn each other of dangers, and even support one another in times of stress.

At first glance, the world of plants seems one of fierce competition. Each tree, each flower, vying for sunlight, water, and space. But what if we told you that plants also engage in deep cooperation? That they not only survive in isolation but thrive because of their social interactions with one another? The roots of connection run far deeper than we might imagine, and in this hidden world, plants create a social fabric that binds them together, nurturing and sustaining their communities.

8.1. Roots as Communicators – The Hidden Language Below

We know that roots anchor plants to the ground, providing them with the stability needed to grow and access nutrients from the soil. However, recent scientific discoveries have unveiled a more profound role for roots: they act as channels of communication and cooperation between plants. These subterranean signals, passed through the soil, connect plants in ways that suggest a far more complex and cooperative existence than previously thought.

One of the most remarkable discoveries in this field is the presence of mycorrhizal networks—often referred to as the "Wood Wide Web." These networks of fungal threads, which intertwine with plant roots, serve as conduits for the exchange of nutrients, water, and even chemical signals between plants. Through these underground networks, plants can "speak" to one another, warning of pest attacks, sharing resources, and even aiding in the survival of neighboring plants.

In one famous experiment, scientists found that when a plant was attacked by aphids, nearby plants began to release volatile organic compounds (VOCs)—chemical signals that act like a distress call. These VOCs alerted surrounding plants to the threat, prompting them to activate their defenses. Through this chemical language, plants could warn each other, creating a kind of early warning system for their community. The ability to communicate through the release of VOCs is not limited to pest defense; plants also use these chemicals to attract pollinators, repel herbivores, and even promote the growth of neighboring plants.

This underground web of communication suggests that plants are not isolated entities, but part of a larger, interdependent community. Through the mycorrhizal network, they share not just resources but information, ensuring the survival and prosperity of the entire ecosystem.

8.2. Cooperative Networks – Plants Helping One Another

While plants do communicate to warn one another of danger, they also engage in forms of cooperation that challenge our traditional understanding of competition in nature. In ecosystems where plants coexist, they often share resources such as water and nutrients through their interconnected roots and fungal networks. In some cases, plants have been shown to redistribute resources—such as carbon and nitrogen—through these underground connections, helping weaker or younger plants survive.

One striking example of this cooperation is found in forests, where older, more established trees often share resources with younger saplings. These mature trees, which have access to greater water and nutrient supplies, "send" extra carbon to the younger plants through their root systems. This seemingly altruistic behavior ensures that the forest ecosystem remains balanced, with each plant benefiting from the collective effort.

Such cooperation is not restricted to trees alone. Many species of plants, including grasses and herbs, have been found to work together in the wild, using their root systems and fungal networks to create a sense of mutual support. In some cases, plants of the same species form "clonal communities," where multiple individuals share a common root system, working in unison to acquire nutrients, water, and sunlight.

The idea of cooperation in plant communities is not only limited to the exchange of resources but extends to the social bonds formed within these networks. In a way, plants "care" for one another, maintaining the health of the ecosystem through their interconnectedness. These cooperative behaviors allow plants to thrive in environments where individual survival would be difficult, creating a web of interdependence that is essential to the overall health of the community.

8.3. The Role of Allelopathy – Plants as Chemical Communicators

In addition to physical networks like roots and fungi, plants also communicate through the release of chemicals into the soil, a phenomenon known as allelopathy. This chemical warfare, often used by plants to protect themselves from competing species, can also play a role in creating social bonds within plant communities.

Through allelopathy, plants release chemicals that inhibit the growth of nearby competitors, ensuring they have access to the resources they need. These chemicals can be toxic to other plants, acting as a defense mechanism against aggressive invaders. However, this chemical communication can also be used in more subtle ways. Some plants release allelopathic compounds that help to regulate the growth of nearby species, ensuring that no one species dominates the area and that a diverse plant community can flourish.

In some cases, plants use allelopathy to support one another. For example, certain species of trees release chemicals that discourage the growth of invasive species, allowing native plants to thrive. In this way, plants can foster a balanced ecosystem, where the release of chemicals is a tool for both defense and cooperation.

This chemical communication extends beyond the plant's immediate surroundings. Research has shown that plants can "remember" past interactions with their environment and adjust their behavior accordingly. If a plant has been exposed to high levels of allelopathic chemicals from another species, it may alter its growth strategy, growing more slowly or altering its root structure to avoid chemical inhibition.

8.4. Plant "Memory" – Long-Term Social Bonds

It is in this chemical and physical communication that we also encounter another fascinating aspect of plant interaction: plant memory. Just as humans form bonds based on shared experiences, plants too can "remember" their past interactions with their environment. This form of memory is less about conscious recollection and more about physiological adaptation.

Plants can "remember" past experiences of drought, pest attacks, or competition for resources, and they adjust their future growth strategies accordingly. This memory is not stored in the same way it is in animals or humans, but it exists as a form of adaptive response to the environment. For example, plants that have previously been exposed to herbivory may develop stronger defenses in subsequent growing seasons, as if they "remember" the danger they once faced.

This long-term memory is not just an individual trait but a collective one. As plants communicate with one another, they share this memory, creating a sense of continuity within the ecosystem. When a plant responds to a pest attack, it not only protects itself but may also warn others through chemical signals or physical cues. In this way, the plant community "remembers" the past together, reinforcing the social bonds that sustain them.

8.5. Conclusion: The Living Web – Plants as Social Beings

The world of plant communication is a testament to the complexity and richness of life that exists beneath the surface of our everyday perceptions. Far from being passive, solitary organisms, plants are deeply connected to one another, forming intricate networks of cooperation and mutual support. Through their roots, their chemical signals, and their physical interactions, plants communicate in ways that reveal a hidden social structure within the natural world.

In the soil, plants are not isolated individuals but part of a vast, interconnected web of life. They share resources, protect one another, and work together to create a balanced ecosystem. These connections form the roots of a plant society, where cooperation is as vital to survival as competition.

As we begin to understand the depth of plant communication, we are reminded of the intricate, often invisible threads that bind all life together. Plants, in their quiet and patient way, have much to teach us about the power of connection, cooperation, and community in the world.

Chapter 9
The Wisdom of the Forest – How Plants Adapt to Their Environment

"The forest is a living story, written in the patterns of growth and survival, etched into the bark of ancient trees and whispered through the leaves in the wind. Each plant, in its quest for light, water, and space, tells a tale of adaptation—a saga not only of existence but of resilience and transformation."

Introduction

The world of plants, though silent, speaks volumes through the intricate ways in which they adapt to their surroundings. A plant's life is a story of constant change, a dance between survival and the desire to flourish. It faces myriad challenges—drought, extreme temperatures, pests, and competition—yet, through an unparalleled mastery of adaptation, it thrives in even the harshest environments. This is the wisdom of the forest, written in the leaves, etched in the bark, and whispered in the wind.

In their pursuit of survival, plants have developed astonishing strategies that allow them to not only endure but to flourish under difficult conditions. They have learned to respond to their environment with a remarkable intelligence, albeit one that operates outside of our human understanding of cognition. This wisdom is woven into the very fabric of their existence—the way they grow, the way they interact with their surroundings, and the way they evolve over time. The adaptability of plants is a form of communication, a language of resilience that speaks directly to their environment.

9.1. Photosynthesis and the Search for Light – A Mastery of Sunlight

At the most basic level, the story of a plant's adaptation begins with the search for sunlight. Like all green plants, trees and shrubs harness the power of photosynthesis to create energy from light. Yet, the way in which plants adapt to the availability of light—how they seek it, how they share it, and how they adjust to its absence—is a tale of profound ingenuity.

Plants, particularly those living in dense forests or shaded canopies, have developed a number of adaptations to make the most of the light they receive. Some species, like the towering trees in a tropical rainforest, grow tall with broad leaves, reaching up and out to capture the sunlight above the forest floor. These trees are like the ancient sentinels of the forest, their roots reaching deep into the earth, while their leaves stretch to the heavens.

Other plants, particularly those in the understory of the forest, have adapted to low-light conditions by evolving large, broad leaves that maximize their ability to absorb light. These plants don't compete for height but have perfected the art of catching the fleeting beams of sunlight that filter down through the canopy. Their leaves, thinner or waxier in texture, are designed to conserve water and energy, ensuring that every ray of light is harnessed to its fullest potential.

In the desert, where sunlight is unrelenting, plants like cacti have adapted by evolving a different strategy altogether. Instead of seeking out sunlight, they store it, capturing and holding onto the energy through specialized cells that allow them to thrive in harsh conditions. Their leaves are modified into spines, which minimize water loss, while their stems swell with water reserves, capable of withstanding long periods of drought. The desert is a realm of extremes, and in it, plants like the cactus have mastered the art of surviving with less, thriving where others would perish.

9.2. Water as a Resource – Adaptations to Arid Environments

Water is perhaps the most precious resource for all living things, and for plants, it is both a source of life and a potential cause of death. The ability to adapt to water availability has shaped the evolution of plant species across the globe, from the lush wetlands to the parched deserts.

In temperate and tropical climates, plants are more accustomed to regular rainfall and access to ample water. However, even within these regions, plants have developed unique mechanisms to manage their water needs. The thick, waxy cuticles of many leaves reduce water loss, while specialized stomata (pores) control transpiration, allowing plants to conserve moisture when conditions are dry. Some plants, like the eucalyptus tree, have adapted to seasonal droughts by shedding leaves during the driest months, reducing their surface area and, consequently, their water loss.

In arid environments, where water is scarce and the sun beats down relentlessly, plants have developed extraordinary strategies for conserving and utilizing water. Succulents, for example, store water in their thick, fleshy tissues, while others like the agave use their deep roots to access groundwater far below the surface. Many desert plants are also equipped with the ability to open their stomata at night when the air is cooler, reducing water loss to the dry, scorching air of the day.

The wisdom of plants in arid regions is truly remarkable. Some species of trees, like the mesquite, have developed long taproots that can reach water sources buried deep beneath the earth's surface. Other plants, like the creosote bush, release chemicals from their roots that inhibit the growth of nearby plants, reducing competition for the limited water in the environment. It is a harsh world, yet these plants have adapted to live in a balance of efficiency, conserving every drop of water they can.

9.3. The Dance with Pests and Disease – Chemical Warfare and Defenses

Beyond the challenges of light and water, plants face another battle—the constant threat of herbivores and pathogens that seek to devour or destroy them. Yet, even in the face of these threats, plants have developed sophisticated defenses, not only to ward off attack but to deter future threats. Their adaptation to these dangers is an arms race of sorts, with plants evolving new chemical and physical defenses to protect themselves.

In response to herbivores, many plants have evolved thorny, spiny, or leathery structures that make them difficult or unpleasant to eat. Others produce toxic compounds that make their leaves or stems poisonous to animals. The alkaloids and terpenoids found in some plants act as potent deterrents to grazers, while others release volatile organic compounds (VOCs) that signal to nearby plants, warning them of the presence of herbivores.

Plants can also form alliances with other species to combat pests. For example, some plants attract predatory insects, such as ladybugs or lacewings, by releasing specific scents that signal the presence of aphids or other pests. In return, these predators devour the herbivores, effectively protecting the plant. Similarly, plants have been found to harbor beneficial fungi that help them resist fungal infections, forming an alliance in the battle against pathogens.

The plant world is one of constant adaptation and response to changing environmental pressures. As herbivores evolve new methods of attacking plants, plants, in turn, adapt with new methods of defense. It is a dynamic relationship, driven by the struggle for survival.

9.4. The Role of Symbiosis in Survival

Perhaps the most profound adaptation plants have developed is their ability to form symbiotic relationships with other organisms. These mutualistic partnerships, where both species benefit, are crucial to the survival of plants in many ecosystems. One of the most important of these relationships is that of plants with their fungal partners, particularly mycorrhizal fungi.

Mycorrhizae form networks with plant roots, extending them into the soil far beyond where the plant's roots could reach alone. In exchange for nutrients such as nitrogen and phosphorus, plants provide the fungi with sugars produced through photosynthesis. This relationship benefits both partners, increasing the plant's access to nutrients while aiding the fungi in the search for food. In forests, this connection is even more profound, with mycorrhizal networks allowing trees to share resources and communicate over vast distances.

Similarly, plants also form relationships with pollinators, such as bees, birds, and bats. These relationships are not just beneficial; they are essential for the reproduction of many plant species. Flowers evolve to attract specific pollinators with unique colors, scents, and shapes, ensuring the plants' reproductive success. In return, the pollinators receive nectar, pollen, or other rewards.

Symbiosis, at its core, is an adaptation that allows plants to thrive in an interconnected world. The wisdom of plants lies in their understanding that cooperation is often as important as competition in the natural world.

9.5. Conclusion: A Legacy of Adaptation and Resilience

The forest is a living testament to the power of adaptation. Each plant, from the tallest oak to the smallest fern, has honed its ability to survive and flourish in its environment. Whether it is through the search for light, the quest for water, or the constant battle against pests and disease, plants have mastered the art of resilience. Their wisdom, passed down through generations, is written in the ways they respond to their environment and interact with the world around them.

In the forest, survival is not just about being the strongest or the fastest—it is about adapting, changing, and finding a way to live in harmony with the environment. Plants have learned to thrive not in isolation, but through cooperation, symbiosis, and communication. They offer us a lesson in resilience, in the power of quiet strength, and in the importance of adapting to the world, rather than resisting it.

The forest, with all its plants and trees, is more than just a collection of life. It is a living, breathing narrative—a story of struggle, survival, and, ultimately, wisdom.

Chapter 10

The Future of Plant Wisdom – How We Can Learn from Plants to Heal the Earth

"The roots of the earth are not only in the soil but in the wisdom that grows within the leaves of the trees. To heal the planet, we need only to listen—to the quiet murmur of roots in the earth, to the rustling whispers of leaves, and to the ancient stories of survival written in every leaf and petal."

Introduction

In the quiet solitude of nature, beneath the deep canopy of leaves and in the tangled roots that weave through the earth, lies a wisdom older than humanity itself. Plants, with their silent resilience, their intricate adaptations, and their ability to live in harmony with the earth, hold lessons that could guide us toward a more sustainable future. As we face the immense challenges of climate change, ecological destruction, and biodiversity loss, the future of our planet may very well depend on our ability to understand—and to learn—from the world of plants.

Plants, in their simplicity and complexity, offer us blueprints for living in harmony with the earth. Their wisdom is not the kind of knowledge that is written down in books or recorded in data—rather, it is embedded in the very processes of life, growth, and survival. From the way they share nutrients with one another, to the way they adapt to extreme conditions, plants have developed strategies for thriving in a world that constantly changes. It is this adaptive intelligence, this quiet resilience, that holds the potential for healing the earth.

10.1. The Power of Regeneration – Plants as Healers of the Earth

One of the most profound aspects of plant life is their ability to regenerate and heal. Unlike many animals, plants can recover from injury, disease, and even catastrophic damage. This regenerative capacity is more than just a survival mechanism—it is a fundamental aspect of the ecological processes that sustain life on earth. It is through their ability to heal and regenerate that plants play a crucial role in restoring ecosystems that have been damaged by human activity.

For instance, after a forest fire or a flood, the earth may seem barren, but plants are among the first to return, their seeds lying dormant, waiting for the right moment to sprout. Through a complex process of ecological succession, plants help to restore the balance of the ecosystem, rebuilding the soil, filtering the air and water, and providing habitats for countless other species. In this sense, plants are not just survivors—they are healers.

In the context of large-scale environmental restoration, the regenerative power of plants can be harnessed to restore degraded landscapes. Techniques like reforestation, agroforestry, and the planting of native vegetation can help bring life back to barren lands, sequestering carbon, improving soil quality, and restoring biodiversity. The lessons we learn from plants in these regenerative processes are not just about healing the earth—they are about respecting the deep, cyclical rhythms of nature and working with, rather than against, them.

10.2. Sustainable Agriculture – Working in Tandem with Nature

Perhaps one of the most pressing issues of our time is the way we produce food. Industrial agriculture, with its heavy reliance on synthetic chemicals, monocultures, and fossil fuels, has taken a heavy toll on the planet. But plants, in their infinite wisdom, can guide us toward a more sustainable and harmonious way of farming—one that works in tandem with nature, rather than exploiting it.

Agroecology, a practice that draws on the wisdom of traditional farming methods and modern ecological science, offers a path forward. It emphasizes the importance of biodiversity, soil health, and natural pest control, while reducing the need for artificial inputs like pesticides and fertilizers. By cultivating a diverse range of plants and working with natural processes, farmers can create resilient agricultural systems that not only feed people but also restore ecosystems.

One such example is the practice of companion planting, in which certain plants are grown together because they benefit one another. For instance, the ancient tradition of planting beans alongside corn and squash—a technique known as the "Three Sisters"—not only helps to control pests but also improves soil health. The beans fix nitrogen in the soil, the corn provides structure for the beans to climb, and the squash provides ground cover, preventing weeds. This method, rooted in Indigenous knowledge, shows how plants can work together to create more resilient, sustainable systems.

Moreover, the concept of permaculture—designing agricultural systems that mimic the patterns and relationships found in nature—also draws on the wisdom of plants. In permaculture, plants are selected and arranged in ways that enhance the ecological balance, providing food, shelter, and medicine while regenerating the land. It is a model of agriculture that honors the plants' ability to adapt, regenerate, and support one another, offering a vision of farming that is as much about healing the land as it is about feeding people.

10.3. The Role of Plants in Carbon Sequestration – A Solution to Climate Change

In the battle against climate change, plants are not just passive bystanders—they are active agents of transformation. Through the process of photosynthesis, plants absorb carbon dioxide from the atmosphere, converting it into oxygen and organic matter. This ability to sequester carbon in their tissues plays a critical role in mitigating climate change.

Forests, particularly tropical rainforests, are the planet's largest carbon sinks, absorbing vast amounts of carbon dioxide and storing it in their biomass. Protecting and restoring these forests is one of the most effective ways to combat climate change. But plants aren't limited to forests—they also sequester carbon in grasslands, wetlands, and even in the soils of agricultural fields. The preservation and restoration of plant ecosystems is, therefore, a key strategy in reducing atmospheric carbon levels and stabilizing the global climate.

In addition to carbon sequestration, plants play an important role in regulating the water cycle, stabilizing soil, and preventing erosion. Plants help maintain the balance of ecosystems, ensuring that water is filtered, soil is enriched, and nutrients are cycled through the environment. This interconnectedness is a reminder that the health of our planet depends on the health of its ecosystems, and that plants are at the heart of this balance.

10.4. The Healing Potential of Plants – Medicine from Nature

Beyond their ecological importance, plants also hold immense potential for healing. For thousands of years, humans have turned to plants for medicine, relying on their natural compounds to treat illnesses, ease pain, and improve health. From the willow tree, which provides the active ingredient in aspirin, to the aloe vera plant, which soothes burns, plants have been our companions in healing.

But there is still much to learn from the plant world. Many plants contain bioactive compounds that have yet to be discovered, compounds that could offer new treatments for diseases that currently have no cure. In the rainforest, scientists are racing to uncover the medicinal secrets hidden in its diverse plant species before they are lost forever to deforestation.

The study of plants in medicine, known as ethnobotany, is an area of growing interest. As we face global health crises, from antibiotic resistance to emerging diseases, plants offer a rich repository of untapped healing potential. Their wisdom lies not only in their ability to adapt and survive, but also in their capacity to offer solutions to some of humanity's most pressing health challenges.

10.5. A Future of Harmony – Learning to Live with Plants, Not Exploit Them

The future of the planet depends on our ability to recognize the intrinsic wisdom of plants and to incorporate that wisdom into our ways of living. It is not enough to see plants as mere resources to be exploited for human gain. We must learn to live with them, not above them. The key to healing the earth lies in understanding that we are not separate from nature—we are part of it. The same principles that govern the life of a plant, from its ability to adapt and regenerate to its deep connection with the earth, are the principles that can guide us toward a more sustainable and harmonious future.

To live in harmony with plants, we must learn to respect their role in the ecosystem, to value their contributions to the planet, and to understand that the health of plants is directly tied to our own well-being. The way forward is one of collaboration, not domination—working with nature, not against it. The wisdom of plants can guide us, but only if we are willing to listen and learn.

10.6. Conclusion: A Call to Action – Heed the Wisdom of the Green World

As we stand at the crossroads of ecological collapse and potential renewal, the wisdom of plants offers a path forward. We must listen to their quiet teachings—how they adapt, how they regenerate, how they share and cooperate. In their resilience, their ingenuity, and their humility, plants offer us the lessons we need to heal the earth.

It is not too late to change course, but we must act now. The future of our planet rests not just in the hands of scientists, policymakers, or activists—but in the wisdom of the plants that have quietly and steadfastly shaped the earth for millions of years. By learning from them, we may yet find a way to restore balance and create a future where both humans and plants can thrive together, in harmony, on a healed and thriving earth.

References

Baluška, F., & Mancuso, S. (2009). *Plant neurobiology: A discourse on new and old paradigms*. Springer.

Barlow, M. (2019). *The hidden life of trees: What they feel, how they communicate – Discoveries from a secret world.* Greystone Books.

Barton, C. (2017). *Medicinal plants of the world: A guide to plants and their medicinal uses*. University of Chicago Press.

Calvo, P., & Keijzer, F. (2011). Plants: Adaptive behavior, root-brains, and minimal cognition. *Adaptive Behavior, 19*(3), 155–171.

Calvo, P., & Keijzer, F. (2009). "Cognition in plants." *Interface Focus*, 1(1), 203–211.

Chamovitz, D. (2012). *What a plant knows: A field guide to the senses*. Scientific American/Farrar, Straus and Giroux.

Gagliano, M. (2017). *Thus spoke the plant: A remarkable journey of ground-breaking scientific discoveries and personal encounters with plants*. North Atlantic Books.

Gagliano, M., Ryan, J., & Vieira, P. (Eds.). (2017). *The Language of Plants: Science, Philosophy, Literature*. University of Minnesota Press.

Griffiths, D. (2015). *The role of forests in the Earth's climate regulation*. Earthscan Publications.

Hawken, P. (2017). *Drawdown: The most comprehensive plan ever proposed to reverse global warming*. Penguin Books.

Hughes, C. (2021). *Regenerative agriculture: A practical guide to working with nature*. EarthWorks.

Jones, C. (2018). *Symbiosis: The nature of cooperative interactions in ecology*. Oxford University Press.

Karban, R. (2015). *Plant Sensing and Communication*. University of Chicago Press.

Klein, R. (2016). *The wisdom of plants: How plants adapt and survive*. Cambridge University Press.

Mancuso, S., & Viola, A. (2015). *Brilliant green: The surprising history and science of plant intelligence*. Island Press.

Meyer, H. (2020). *Agroecology: The ecology of sustainable food systems*. CRC Press.

Parker, M. (2016). *Pollination: A forgotten resource*. Springer International.

Pollan, M. (2001). *The botany of desire: A plant's-eye view of the world*. Random House.

Sachs, J. D. (2020). *The age of sustainable development*. Columbia University Press.

Seneff, S., & Albright, L. (2018). *Toxic legacy: How the chemicals we use in agriculture are killing us and our planet*. Greenleaf Publishing.

Sheldrake, M. (2020). *Entangled life: How fungi make our worlds, change our minds & shape our futures*. Random House.

Tilman, D., & Clark, M. (2015). Global food security and agricultural sustainability: A critical review. *Science, 347*(6229), 1237–1243. https://doi.org/10.1126/science.1254732

Trewavas, A. (2014). *Plant behaviour and intelligence*. Oxford University Press.

Vega, F. E., & Glare, T. R. (2020). *Fungi as biocontrol agents: Progress, challenges, and future directions*. Springer.

Wohlleben, P. (2016). *The Hidden Life of Trees*. Greystone Books.

Kolibri 360 GmbH
Ferdinand-Jühlke-Straße 7
99095 Erfurt, Deutschland
produktsicherheit@kolibri360.de

Druck:
CPI Druckdienstleistungen GmbH
im Auftrag der
Kolibri 360 GmbH
Ferdinand-Jühlke-Str. 7
99095 Erfurt